ACE GENERAL CHEMISTRY II

(THE EASY GUIDE TO ACE GENERAL CHEMISTRY II)

BY: DR. HOLDEN HEMSWORTH

Copyright © 2015 by Holden Hemsworth

All rights reserved. No part of this publication may be reproduced, distributed, or transmitted in any form or by any means, including photocopying, recording, or other electronic or mechanical methods, without the prior written permission of the publisher, except in the case of brief quotations embodied in critical reviews and certain other noncommercial uses permitted by copyright law.

DISCLAIMER

Chemistry, like any field of science, is continuously changing and new information continues to be discovered. The author and publisher have reviewed all information in this book with resources believed to be reliable and accurate and have made every effort to provide information that is up to date and correct at the time of publication. Despite our best efforts we cannot guarantee that the information contained herein is complete or fully accurate due to the possibility of the discovery of contradictory information in the future and any human error on part of the author, publisher, and any other party involved in the production of this work. The author, publisher, and all other parties involved in this work disclaim all responsibility from any errors contained within this work and from any results that arise from the use of this information. Readers are encouraged to check all information in this book with institutional guidelines, other sources, and up to date information.

The information contained in this book is provided for general information purposes only and does not constitute medical, legal or other professional advice on any subject matter. The information author or publisher of this book does not accept any responsibility for any loss which may arise from reliance on information contained within this book or on any associated websites or blogs.

WHY I CREATED THIS STUDY GUIDE

In this book I try to breakdown the content covered in the second semester of a typical General Chemistry course in college for easy understanding and to point out the most important subject matter that students are likely to encounter. This book is meant to be a supplemental resource to lecture notes and textbooks to boost your learning and go hand in hand with your studying!

I am committed to providing my readers with books that contain concise and accurate information and I am committed to providing them tremendous value for their time and money.

Best regards,
Dr. Holden Hemsworth

Table of Contents

CHAPTER 1: Review of Fundamental Concepts 1

CHAPTER 2: Gases and Gas Laws 13

CHAPTER 3: Thermochemistry 16

CHAPTER 4: Solutions 23

CHAPTER 5: Chemical Kinetics 28

CHAPTER 6: Chemical Equilibrium 34

CHAPTER 7: Acid Base Equilibrium 37

CHAPTER 8: Solubility Equilibrium 44

CHAPTER 9: Electrochemistry 47

CHAPTER 10: Nuclear Chemistry 52

CHAPTER 1 – REVIEW OF FUNDAMENTAL CONCEPTS

Matter

Matter is anything that has mass and takes up space. Mass is the amount of matter an object contains; a way of quantifying matter. Matter exists in three physical states.

- Solid – matter with fixed shape and volume (rigid)
- Liquid – matter with a fixed volume but indefinite shape
    - Takes on the shape of the container it is in
- Gas – matter without a fixed shape or volume
    - Conforms to the volume and shape of its container

Components of Matter (Definitions)

- Element - substance that can't be broken down into other substances by chemical means
- Molecule - a combination of two or more atoms
- Compound – substance formed from two or more chemical elements that are chemically bonded together
- Mixture - two or more elements (or compounds) mingling without any chemical bonding

Physical and Chemical Properties

- Physical property – characteristics that can be measured and observed without changing the chemical makeup of the substance
    - Examples: color, melting point, boiling point, density, etc.
- Physical change – a substance changes its physical appearance but does not change identity
    - Changes in state (e.g., liquid to gas, solid to liquid) are all physical changes
- Chemical property – any property that becomes evident during a chemical reaction
    - Examples: pH, corrosiveness, etc.

- Chemical change (aka chemical reactions) – a substance is transformed into a chemically different substance

Laws of Matter

- Law of Mass Conservation
    - Total masses of substances involved in a chemical reaction do not change
        - Number of substances and their properties can change
- Law of Definite Proportions:
    - Pure compounds contain exactly the same proportions of elements by mass
- Law of Multiple Proportions
    - If two elements react to form more than one compound, then the ratios of the masses of the second element which combine with a fixed mass of the first element will be in ratios of small whole numbers

Periodic Table of Elements

The periodic table is an arrangement of elements in rows and columns based on their atomic number, electron configurations, and chemical properties.

- Period – horizontal row on the table
- Group (Family) – column on the table

- Elements on the periodic table can be classified as metals, nonmetals, and metalloids
    - Metal – substances that have luster, high heat conductivity, high electrical conductivity, and are solid at room temperature (exception: mercury)
    - Nonmetal – substance without any metal characteristics
    - Metalloid – substance that have both metal and nonmetal characteristics

Atoms

An atom is the smallest unit of matter. Atoms interact to form molecules. Atoms are composed of subatomic particles (electrons, protons, and neutrons).

- Electrons – negatively charged particles
    - Carries a charge of -1.602×10^{-19} Coulombs (C)
        - Charge of atomic and sub-atomic particles are typically described as a multiple of this value
        - So, referred to as -1
    - Mass = $9.10938291 \times 10^{-31}$ kg
- Protons – positively charged particles
    - Carries a charge of $+1.602 \times 10^{-19}$ Coulombs (C)
        - Referred to as a +1 electron charge
    - Mass = $1.67262178 \times 10^{-27}$ kg
- Neutrons – uncharged particles
    - Electrically neutral
    - Mass = $1.674927351 \times 10^{-27}$ kg
- Protons and neutrons are found in the nucleus (central core of an atom)
    - Electrons orbit the nucleus in an "electron cloud"
- Elemental (atomic) symbol: shorthand representation of atoms of different elements

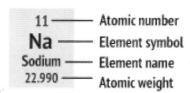

Example of an Element on the Periodic Table:

- Atomic number - number of protons in an atom of a particular element
    - For a neutral atom, number of electrons = number of protons

- o All atoms of an element have the same atomic number (same number of protons)
- Mass number = the number of protons + the number of neutrons
  - o All atoms of an elements don't have the same number of neutrons
- Atomic weight (relative atomic mass) – average mass of atoms of an element
  - o Calculated based on the relative abundance of isotopes in that particular element
  - o Units: atomic mass units (amu)
- Isotopes – atoms of an element with the same number of protons but with a different number of neutrons
  - o Same atomic mass but different mass number

Types of Chemical Formulas

Chemical formulas are a way of expressing information about the proportions of atoms that constitute a compound using: element symbols, numerical subscripts, and other symbols (e.g., parentheses, dashes).

- Empirical formula – smallest whole number ratio of numbers of the atoms in a molecule
- Molecular formula – actual number of atoms in a molecule
- Structural formula – chemical formula showing how atoms are bonded together in a molecule

Ions

Ions are charged atoms or molecules. Ions are formed when atoms or groups of atoms gain or lose valence electrons.

- Monatomic ion – single atom with more or less electrons than the number of electrons in the atom's neutral state
- Polyatomic ions – group of atoms with excess or deficient number of electrons
- Anion – negatively charged ion
- Cation – positively charged ion
- Ionic compounds – association of a cation and an anion
  - o The cation is always named first

Chemical Equations

- Chemical reactions are expressed through chemical equations
- An arrow ("→") in a chemical equation means "yields"
    - $2 H_2(g) + O_2(g) \rightarrow 2 H_2O(l)$
        - Hydrogen + oxygen yields water
    - H_2 and O_2 are reactants
        - Substances that undergo change during a reaction
    - H_2O is the product
        - Substances formed from chemical reactions
- Common phase notation
    - g = gas
    - l = liquid
    - s = solid

Balancing Chemical Equations

- Balanced chemical equations adhere to the Law of Conservation of Matter
    - A balanced equation has to have equal numbers of each type of atom on both sides of the arrow
- Balancing is done by changing the coefficients
    - The coefficient times the subscript gives the total number of atoms
    - If there are no coefficients in front, coefficient is equal to one
    - If an atom doesn't have a subscript, subscript is equal to one
- Subscripts are **never** changed

Periodic Properties

Periodic Law states that when elements are arranged by atomic number, their physical and chemical properties vary across the periodic table row.

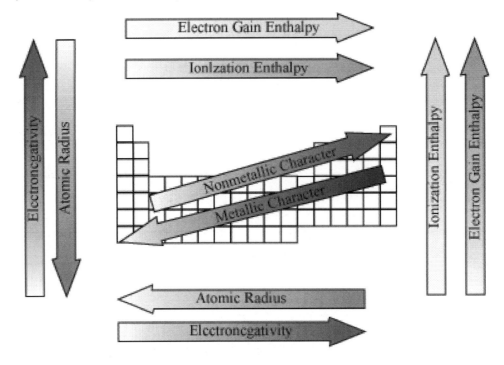

Atomic Size

- Two factors that affect size of an atom
    - Larger the principal quantum number (n), the larger the size of the orbit
    - Effective nuclear charge
        - Positive charge an electron experiences from the nucleus minus any shielding effects
- Atomic radius tends to **decrease** with increasing atomic number across a period
    - There is a higher effective nuclear charge
        - Greater attractive force in the nucleus from the higher number of protons
- Atomic radius tends to **increase** with increasing atomic number within a group
    - More energy levels
        - Each subsequent energy level is further from the nucleus than the last

Ionic Size

- Cations are **smaller** than their neutral atom counterparts
  - Electrons are removed
    - Results in a decrease in electron repulsion
      - Allows nuclear charge to pull electrons closer
- Anions are **bigger** than their neutral atom counterparts
  - Electrons are added
    - Results in an increase in electron repulsion
      - Occupy more space
- Ionic size **increases** down a group
  - More energy levels
- Ionic size is slightly complicated across a period
  - Decreases among cations
  - Increases dramatically with the first anion
    - Decreases within anions

Ionization Energy (IE)

- First ionization energy - the minimal energy needed to remove one of the outermost electrons from a neutral atom
  - Successive electron removal is called second ionization energy, third ionization energy, etc.
  - Successive ionization energies increase because each electron that is pulled away creates a larger positive charge (i.e. a higher effective nuclear charge)
- Ionization energies tend **increase** with atomic number within a period
  - More difficult to remove an electron that is closer to the nucleus
    - Remember: There is a higher effective nuclear charge across a period
- Ionization energies tend to **decrease** down a group

Electron Affinity

This is not the same as electronegativity!

- More negative electron affinity value express that a more stable negative ion is formed
    - Negative values indicate that energy is released when the anion for that element forms
- General trend is that values become more negative from lower left to upper right
    - Highest electron affinities occur for F and Cl

Electronegativity

- Electronegativity is the measure of an atom's ability of to draw bonding electrons to itself in a molecule
- Electronegativity tends to increase from the lower-left corner to the upper-right corner of the periodic table

Ionic Bonds

Ionic bonds are chemical bonds formed by the electrostatic attraction between positive and negative ions.

- Ionic bonds involve the transfer of electrons from one atom to another
    - Usually the transfer is from a metal from Group IA or IIA to a nonmetal from Group 7A or the top of Group 6A
- Number of electrons lost or gained by an atom is determined by its need to be isoelectronic with its nearest noble gas
    - Noble gas configurations are extremely stable
- Ionic bonds result in the formation of ions that are electrostatically attracted to one another
    - Anion – negatively charged ion
        - Size of an anion is larger than the original size of the neutral atom
    - Cation – positively charged ion
        - Size of a cation is smaller than the original size of the neutral atom

Properties of Ionic Compounds

- Hard (don't dent), rigid (don't bend), brittle (crack but don't deform)
- The above properties are a result of the powerful attractive forces holding ions together

Covalent Bonds

- Two nonmetals often form covalent bonds
    - Share electrons since they have similar attractions for them
- Each nonmetal holds tightly to its own electrons
    - Shared electron pair spend most of their time between the two atoms
        - Electron pairs being shared are said to be localized
- How covalent bonds form
    - Distance between two nuclei decreases
        - Each starts to attract the other's electron(s)
        - Causes a decrease in potential energy
    - Atoms draw closer and closer together
        - Energy becomes progressively lower
    - Attractions increase but so does repulsions between electrons
    - At a particular internuclear distance, maximum attraction is achieved
        - Balance between nucleus-electron attractions
        - Balance between electron-electron and nucleus-nucleus repulsions
- Two sets of forces are involved within covalent compounds
    - Strong covalent bonding forces hold atoms together within the molecule
    - Weak intermolecular forces hold separate molecules near each other
        - Weak intermolecular forces between molecules are responsible for the observed physical properties of these molecules

Types of Covalent Bonds

- Coordinate covalent bond – covalent bond in which both of the shared electrons are donated by a single atom
- Double bond – sharing of two pairs of electrons between atoms
- Triple bonds – sharing of three pairs of electrons between atoms
- Polar covalent bond – covalent bond where the bonding electron spend more time closer to one of the atoms involved in the bonding

- Nonpolar covalent bond – covalent bond where the bonding electrons are shared equally

Bonding Pairs and Lone Pairs

- Shared electrons are considered as belonging entirely to each atom in a covalent bond
    - Shared electron pair simultaneously fills the outer level of both atoms
- An outer-level electron pair that is not involved in bonding is a "lone pair"

Metal-Metal Bonding

- Metals have low ionization energies
    - Lose electrons easily
    - Do not gain them readily
- Valence electrons are evenly distributed around metal-ion cores
    - Metal-ion cores consist of the nucleus plus the inner electrons
- Valence electrons are delocalized
    - Move freely throughout the metal

Predicting Ionic and Covalent Bonding

- Non-polar covalent bond
    - Typically electronegativity difference between the two atoms has to be less than 0.5 for non-polar bonds
- Polar covalent bonds
    - Electronegativity between the two atoms is different by a greater degree than 0.5 but less than 2.0
- Ionic bonds
    - Typically, difference in electronegativity is more than 2.0

Octet Rule

The octet rule states that the tendency of atoms in a molecule is to have eight electrons in their outer shell.

- There are exceptions to this rule where the central atom may have more than eight electrons
- Generally, a nonmetal in the third period or higher can accommodate as many as twelve electrons, if it is the central atom
    - These elements have unfilled "d" subshells that can be used for bonding

Resonance (Delocalized Bonding)

- Structures of some molecules can be represented by more than one Lewis dot formula
    - Individual Lewis structures are called contributing structures
    - Individual contributing structures are connected by double-headed arrows (aka resonance arrows)
    - Molecule or ion is a hybrid of the contributing structures and displays delocalized bonding
        - Delocalized bonding is where a bonding pair of electrons is spread over a number of atoms
- Some resonance structures contribute more to the overall structure than others
    - Determining which structures are more contributing
        - Structures where all atoms have filled valence shells
        - Structures with the greater number of covalent bonds
        - Structures with less charges
            - Formal charges can help discern which structure is most likely (discussed later in this section)
        - Structures that carry a negative charge on the more electronegative atom

Example of Resonance Structures:

- Curved arrow – symbol used to the redistribution of valence electrons
  - Always drawn as noted in the figure below

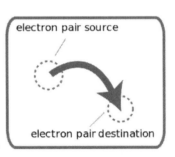

How Curved Arrows are Drawn:

Chapter 2 - Gases and Gas Laws

Properties of Gases

- Flow freely
- Relatively low densities
- Gases form homogenous mixtures with each other
- Volume of gases change significantly with pressure
  - Volumes of solids and liquids volumes are not greatly affected by pressure
- Volume of gases changes significantly with temperature
  - Under high temperatures gases expand
  - Under low temperatures gases contract

Pressure

- Force (F) - a function of the mass of an object under acceleration
  - F = Mass x Acceleration
- Pressure (P) - force exerted per unit area of surface by molecules in motion
  - $P = \dfrac{Force}{Area} = \dfrac{ma}{A} = \dfrac{mg}{A} = \dfrac{mg}{\frac{v}{h}} = \dfrac{mgh}{v} = dgh$
    - A = area
    - m = mass
    - a = acceleration
    - d = density
    - g = acceleration due to gravity = 9.81 m/s^2
    - h = height of column (m)
  - Pressure has many units
    - SI unit is Pascal (Pa) = 1 kg•m^{-1}•sec^{-2}
    - Atmosphere and torr are commonly used
    - 1 atmosphere (atm) = 760 mm Hg = 760 torr

Gas Laws

The behavior of gas can be described by pressure (P), temperature (T), volume (V), and molar amount (n). If you hold any of the two variables constant, it allows for determination of a relationship between the other two.

- Ideal gas – gas that exhibits linear relationships among pressure, temperature, volume, and molar amount
    - Ideal gases don't actually exist
        - But simple gases behave ideally under normal temperatures and pressures
- Molar gas volume (V_m) – volume of one mole of gas
- Volumes of gases are often compared at standard temperature and pressure (STP)
    - Standard Temperature = 0°C (273 K)
    - Standard Pressure = 1 atm
    - V_m at STP = 22.4 L/mol
- Boyles Law
    - Volume of a sample of gas at a constant temperature is inversely related to the applied pressure
    - $P_1V_1 = P_2V_2$ or $\frac{P_1}{P_2} = \frac{V_1}{V_2}$
- Charles Law
    - Volume of a sample of gas at constant pressure is directly proportional to the absolute temperature
    - $\frac{V_1}{T_1} = \frac{V_2}{T_2}$
- Avogadro's Law
    - Equal volumes of different gases at the same temperature and pressure contain equal number of particles
    - $\frac{V_1}{n_1} = \frac{V_2}{n_2}$
- Combined Gas Law
    - For when P, V, and T are changing
        - $\frac{P_1V_1}{T_1} = \frac{P_2V_2}{T_2}$

- Ideal Gas Law
    - PV = nRT
        - R is the universal gas constant
        - R = 0.082058 L•atm•mol^{-1}•K^{-1} = 8.3145 J•mol^{-1}•K^{-1}
    - As long as you know three of the variables you can manipulate the ideal gas law to solve for the fourth
- Molar Mass from Ideal Gas Law
    - $M = \frac{mRT}{PV}$
- Density
    - $d = \frac{P(MM)}{RT}$
- Law of Partial Pressures
    - $P_i = \frac{n_i}{n_{tot}} P_{tot} = x_i P_{tot}$
        - $x_i = n_i / n_{tot}$

Kinetic-Molecular Theory of Gases

- A model based on actions of individual atoms
    - Gases consist of particles in constant motion
    - Pressure derived from bombardment with container
    - Kinetic energy formula: $E_k = ½ mv^2$
- Postulates of Kinetic Theory
    - Volume of particles is negligible
    - Particles are in constant motion
    - No inherent attractive or repulsive forces
    - The average kinetic energy of a collection of particles is proportional to the temperature (K)
- Molecular Motion in Gases
    - Diffusion – transfer of gas through space or another gas over time
    - Effusion – transfer of a gas from a region of high pressure to a region of low pressure

CHAPTER 3 - THERMOCHEMISTRY

Thermochemistry

- In chemical reactions whenever matter changes composition, its energy content changes as well
- In some reactions the energy contained in the reactants is higher than the energy contained in the products
    - The excess energy is released as heat
- In other reactions the energy contained in the reactants is lower than the energy contained in the products
    - In these reactions, energy (heat) must be added before the reaction can proceed
- Physical changes also involve a change in energy
- Thermodynamics - science of the relationship between heat and other forms of energy
- Thermochemistry - study of the quantity of heat absorbed or exuded by chemical reactions

Energy

Energy is the potential or capacity to do work. Energy is a property of matter and comes in many forms.

- Forms of energy
    - Radiant energy - electromagnetic radiation
    - Thermal energy - associated with random motion of a molecule or atom
    - Chemical energy - energy stored within the structural limits of a molecule or atom
- Concepts of Energy
    - Kinetic energy (E_k) – energy possessed by an object due to its motion
        - $E_k = \frac{1}{2} mv^2$
    - Potential energy (E_p) – energy stored in matter because of its position or location
        - $E_p = mgh$
    - Internal energy (E_i or U_i) – energy associated with the random disordered motion of molecules
        - $E_i = E_k + E_p$

- Units of Energy
    - SI unit of energy is the Joule (J) = kg·m^2/s^2
    - Calorie (cal) - amount of heat required to raise the temperature of one gram of water by one degree Celsius
        - 1 cal = 4.181 J
- When reactants interact to form products and the products are allowed to return to the starting temperature, the Internal Energy (E) has changed
    - ΔE = change in energy
        - Δ is the symbol for change
            - Δ = final value – initial value
    - ΔE = E_{final} - $E_{initial}$ = $E_{products}$ - $E_{reactants}$
- If energy is lost to the surroundings
    - E_{final} < $E_{initial}$
    - ΔE < 0
- If energy is gained from the surroundings
    - E_{final} > $E_{initial}$
    - ΔE > 0

Heat of Reaction

In chemical reactions, heat is either transferred from the "system" to its "surroundings," or vice versa.

- Thermodynamic system - quantity of matter or the space under thermodynamic study
- Surroundings - everything in the vicinity of the thermodynamic system that interacts with the system
- Heat (q) - energy that flows into or out of a system because of a difference in temperature between the system and its surroundings
    - Heat flows from a region of higher temperature to a region of lower temperature
        - Once temperatures equalize, heat flow stops
    - When heat is released from the system to the surrounding
        - q < 0
        - Reaction is called an **exothermic reaction**

- When heat is absorbed from the surrounding by the system
    - $q > 0$
    - Reaction is called an **endothermic reaction**
- Heat of reaction - the value of "q" required to return a system to a given temperature when the reaction goes to completion

Work

Internal energy is specifically defined as the capacity of a system to do work.

- Work – energy transferred when an object is moved by a force
    - $w = -P\Delta V$
    - $\Delta E = q_p + w$
    - $\Delta E = q_p + -P\Delta V$
    - $q_p = \Delta E + P\Delta V$
        - q_p - heat absorbed from the surroundings by the system
        - ΔE - change in internal energy
        - ΔV - change in volume
        - P – pressure

Enthalpy and Enthalpy Change

- Enthalpy (H) - an extensive property of a substance that is used to obtain the heat absorbed or exuded in a chemical reaction
    - $H = E + PV$
    - Enthalpy is a state function
        - Property of a system that depends only on its state at the moment and is independent of any history of the system
    - Enthalpy is representative of the heat energy tied up in chemical bonds
- Change in enthalpy (ΔH) - heat added or lost by the system, under constant pressure
    - $\Delta H = \Delta E + P\Delta V$
    - $\Delta H = q_p$
    - Change in enthalpy is also called the enthalpy of reaction
    - $\Delta H_{rxn} = H_{(products)} - H_{(reactants)}$

- If the system has higher enthalpy at the end of the reaction
    - It absorbed heat from the surroundings
    - It is an endothermic reaction
    - $H_{final} > H_{initial}$
    - ΔH is positive ($+\Delta H$)
- If the system has lower enthalpy at the end of the reaction
    - It exuded heat to the surroundings
    - It is an exothermic reaction
    - $H_{final} < H_{initial}$
    - ΔH is negative ($-\Delta H$)

Thermochemical Equation

Thermochemical equations are chemical reaction equations with the enthalpy of reaction (ΔH_{rxn}) written directly after the equation.

- Example of a thermochemical equation
    - $2\ H_{2\ (g)} + O_{2\ (g)} \rightarrow 2\ H_2O_{(l)}$ $\Delta H_{rxn} = -571.6$ kJ
        - The negative value for ΔH_{rxn} is telling you that heat is lost to the surrounding
        - Also that the equation is exothermic
- Rules for manipulating thermochemical equations
    - If the thermochemical equation is multiplied by some factor, the value of ΔH for the new equation is equal to the ΔH in the original equation multiplied by that factor
    - If the chemical equation is reversed, the sign of ΔH must be reversed
        - Example, if you were to reverse the direction of the equation mentioned above you would get:
            - $2\ H_2O_{(l)} \rightarrow 2\ H_{2\ (g)} + O_{2\ (g)}$ $\Delta H_{rxn} = +571.6$ kJ

Measuring Heats of Reaction

- Heat capacity – amount of heat required to raise the temperature of an object or substance
    - Varies between substances

- Molar heat capacity (C) – amount of heat required to raise the temperature of **one mole** of a substance by **one degree Celsius**
    - $q = nC\Delta T$
        - $\Delta T = T_{final} - T_{initial}$
- Specific heat capacity (S) – amount of heat required to raise the temperature of **one gram** of a substance by **one degree Celsius**
    - $q = mS\Delta T$
        - $\Delta T = T_{final} - T_{initial}$
        - Units for S: J/g·°C
        - m = grams of a sample
- Hess's Law of Heat Summation
    - For a chemical equation that can be written as the sum of two or more steps, the enthalpy changes for the individual steps can be summed (added) up to determine the enthalpy change for the overall equation
    - For coupled reactions, the individual enthalpy changes can be summed up to determine the enthalpy change for the overall reaction

Hess's Law Example Question: What is the standard enthalpy of reaction for the reduction of iron (II) oxide by carbon monoxide? $FeO_{(s)} + CO_{(g)} \rightarrow Fe_{(s)} + CO_{2(g)}$

- Given Information:
    - Equation 1: $3\ Fe_2O_{3(s)} + CO_{(g)} \rightarrow 2\ Fe_3O_{4(s)} + CO_{2(g)}$ $\Delta H = -48.26$ kJ
    - Equation 2: $Fe_2O_{3(s)} + 3\ CO_{(g)} \rightarrow 2\ Fe_{(s)} + 3\ CO_{2(g)}$ $\Delta H = -23.44$ kJ
    - Equation 3: $Fe_3O_{4(s)} + CO_{(g)} \rightarrow 3\ FeO_{(s)} + CO_{2(g)}$ $\Delta H = +21.79$ kJ
- Changes have to made to the above equations to equal the equation in the question
    - Reverse equation 3 and multiply it by two
        - Puts FeO on the reactant side and moves 2 Fe_3O_4 to the products
    - Reverse equation 1
        - Puts Fe_3O_4 on opposite side to cancel with the reverse of equation 3
    - Multiply equation 2 by three
        - Gives 3 Fe_2O_3 on the reactants that will be used to cancel

- New equations after changes
    - Equation 1: $2\ Fe_3O_{4(s)} + CO_{2(g)} \rightarrow 3\ Fe_2O_{3(s)} + CO_{(g)}$ ΔH = +48.26 kJ
    - Equation 2: $3\ Fe_2O_{3(s)} + 9\ CO_{(g)} \rightarrow 6\ Fe_{(s)} + 9\ CO_{2(g)}$ ΔH = -70.32 kJ
    - Equation 3: $6\ FeO_{(s)} + 2\ CO_{2(g)} \rightarrow 2\ Fe_3O_{4(s)} + 2\ CO_{(g)}$ ΔH = -43.58 kJ
- Summing the three equations gives
    - $6\ FeO_{(s)} + 6\ CO_{(g)} \rightarrow 6\ Fe_{(s)} + 6\ CO_{2(g)}$ ΔH = -65.64 kJ
- Dividing by six gives the equation in the question and the final answer
    - $FeO_{(s)} + CO_{(g)} \rightarrow Fe_{(s)} + CO_{2(g)}$ ΔH = -10.94 kJ

Standard Enthalpies of Formation

- Standard state refers to the standard thermodynamic conditions
    - Pressure - 1 atm (760 mm Hg)
    - Temperature - 25°C (298 K)
- Enthalpy change for a reaction where reactants are in their standard states is called the "Standard Heat of Reaction"
    - $\Delta H°_{rxn}$
- Standard enthalpy of formation of a substance - enthalpy change for the formation of one mole of a substance in its standard state from its component elements in their standard states
    - Standard enthalpy of formation for a "pure" element (C, Fe, O, N, etc.) in its standard state is zero
- Law of Summation of Heats of Formation
    - The standard heat of reaction ($\Delta H°_{rxn}$) is equal to the total formation energy of the products minus the total formation energy of the reactants
        - $\Delta H°_{rxn} = \sum n\Delta H°_f(products) - \sum m\Delta H°_f(reactants)$
        - m and n are coefficients of the substances in the chemical equation

Gibbs Free Energy

Gibbs free energy can be used to determine the direction of the chemical reaction under given conditions.

- $\Delta G = \Delta H - T\Delta S$ or $\Delta G = G_{products} - G_{reactants}$
    - G = Gibbs free energy (J/mol)
    - H = enthalpy (J/mol) - total energy content of a system
    - S = entropy (J/K*mol) - measure of disorder or randomness (how energy is dispersed)
    - T = Temperature (K)
        - As T increases so does S
- $+\Delta G$ means energy must be put into the system
    - Indicates that a process is **nonspontaneous or endergonic**
    - Indicates that the position of the equilibrium for a reaction favors the products
- $-\Delta G$ means energy is released by the system
    - Indicates that a process is **spontaneous or exergonic**
    - Indicates that the position of the equilibrium for a reaction favors the reactants
- $\Delta G = 0$ indicates that the system is at equilibrium

Important: ΔG only indicates if a process occurs spontaneously or not, but does **not** indicate anything about how fast a process occurs.

CHAPTER 4 - SOLUTIONS

Solutions

- Solutions are composed of a solute and a solvent
    - Solutions are homogeneous mixtures
    - Solutes are minor components in a solution (present in smaller amounts)
    - Solvents are substances in which a solute is dissolved
- Solution composition is expressed by:
    - Mass percent
        - $mass\ \% = \dfrac{mass\ solute}{mass\ of\ solution} \times 100\%$
        - $mass\ \% = \dfrac{mass\ solute}{mass\ solvent\ +\ mass\ solute} \times 100\%$
    - Mole fraction
        - $X_{solute} = \dfrac{mass\ solute}{mass\ solvent\ +\ mass\ solute}$
        - $X_{solute} = n_{solute} / n_{total}$
    - Molarity
        - $M = \dfrac{moles\ solute}{Liters\ of\ solution} = \dfrac{mol}{L}$
    - Molality
        - $m = \dfrac{moles\ solute}{kg\ solvent} = \dfrac{mol}{kg}$

Conversion between Molarity and Molality

Density must be known to convert from molarity to molality directly.

Example Question: What is molality of 2.00 M NaCl$_{(aq)}$ solution with a density of 1.08 g/mL?

- Determine the mass of 2.00 mol of NaCl using the molar mass of NaCl
    - 2.00 mol x (58.5 g /mol) = 117 g NaCl

- Assume you have 1.000L (1000 mL) of the 2.00 M NaCl solution
    - You can assume any amount if it is not explicitly stated in the question

    Hint: it is best to assume an amount that is easy to work with
        - To convert molarity to molality assume 1.000 L of solution
        - To convert molality to molarity assume 1.000 kg of solvent
- Use the density given in the problem to determine the total mass of the solution
    - 1000 mL x (1.08 g/mL) = 1080 g total mass
- Determine the mass of the solvent
    - Earlier we calculated that we had 117 g NaCl and the total mass of the solution was 1080 g
    - So, to determine the mass of the solvent we will simply subtract the difference
        - 1080 g of solution – 117 g NaCl = 963 g solvent (0.963 kg solvent)
- Finally, use the formula for molality to determine the answer
    - $m = \dfrac{moles\ solute}{kg\ solvent} = \dfrac{2.00\ mol}{0.963\ kg} = 2.08\ \dfrac{mol}{kg}$

Concentrated vs. Dilute Solutions

- Concentrated solution – a solution that contains a high amount of solute
    - More solute particles per unit volume
- Dilute solution – a solution that contains a low amount of solute
    - Fewer solute particles per unit volume
- Saturated solution - a solution that contains the maximum amount of solute that can be dissolved by the solvent

Dilutions

- Extra solvent is added to a solution to dilute it
    - The amount of solute in the solution remains the same

- Use the following formula to solve dilution problems:
    - $M_1V_1 = M_2V_2$
        - M_1 – is the concentration of stock solution
        - V_1 – is the volume of stock solution
        - M_2 – is the concentration of the final solution
        - V_2 – is the volume of the final solution

Henry's Law

Formulated by William Henry in 1803, it states: "At a constant temperature, the amount of a given gas that dissolves in a given type and volume of liquid is directly proportional to the partial pressure of that gas in equilibrium with that liquid."

- $C = k \times P_{gas}$
    - C – solubility of a gas at a fixed temperature in a particular solvent
    - k - Henry's Law constant
    - P_{gas} – partial pressure of the gas

Colligative Properties

Colligative properties refer to properties of solutions that depend upon the ratio of the number of solute particles to the number of solvent molecules in a solution. These properties don't depend on the type of chemical species present.

- Common colligative properties
    - Vapor pressure lowering
    - Freezing point depression
    - Boiling point elevation
    - Osmotic pressure (discussed in detail in its own section)

Vapor Pressure Lowering

- Raoult's Law says that if you add a nonvolatile solute to a solvent, you will cause the vapor pressure of the solute to be lower
- Raoult's Law equation:
    - $P_{solution} = X_{solvent} P°_{solvent}$
        - $P_{solution}$ - vapor pressure of solution
        - $X_{solvent}$ - mole fraction of solvent in solution

- P° solvent - vapor pressure of pure solvent
   - Because $X_{solvent}$ is a mole fraction (a number between 0 and 1), $P_{solution}$ is always lower than $P°_{solvent}$

Freezing Point Depression

- Property based on the observation that the freezing points of solutions are lower than that of the pure solvent and is directly proportional to the molality of the solute
- Formula: $\Delta T_f = T_{f(solvent)} - T_{f(solution)} = K_f \times m$
   - ΔT_f - freezing point depression
   - $T_{f(solvent)}$ – freezing point of the solvent
   - $T_{f(solution)}$ – freezing point of the solution
   - K_f = freezing point depression constant
   - m – molality

Boiling Point Elevation

- Property based on the observation that the boiling point of a solvent is higher when another compound is added (i.e. solution has a higher boiling point than a pure solvent)
- Formula: $\Delta T_b = K_b \times m$
   - ΔT_b – boiling point elevation
   - K_b = boiling point elevation constant
   - m – molality

Osmotic Pressure

Semipermeable membranes stop solute molecules or ions from passing through but allow passage of solvent molecules. Solvent molecules such as water will go through membranes to dilute a solution unless a pressure equal to the osmotic pressure is applied to stop the flow. So, osmotic pressure is defined as the minimum pressure which needs to be applied to a solution to prevent the inward flow of water across a semipermeable membrane.

- Hypertonic - refers to a solution that has higher osmotic pressure than a particular fluid (e.g. intracellular fluid)
- Isotonic - refers to a solution that has the same osmotic pressure than a particular fluid (e.g. intracellular fluid)
- Hypotonic - refers to a solution that has the lower osmotic pressure than a particular fluid (e.g. intracellular fluid)

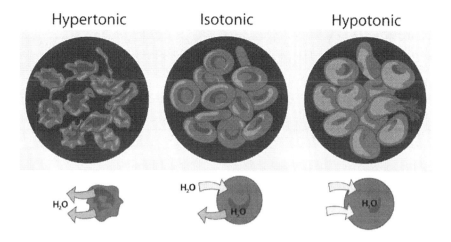

- Osmotic pressure (π) formula
  - πV = nRT or π = MRT
    - n = moles of solute
    - V = volume of solution (L)
    - R = gas constant (0.08206 L·atm/mol·K)
    - T = temperature in Kelvin
    - M = molarity

Hydrophobic Effect and Amphiphilic Molecules

The hydrophobic effect is the tendency of nonpolar substances to aggregate in aqueous solution and exclude water molecules. Amphiphilic molecules have both hydrophobic and hydrophilic parts.

- Hydrophobic – water "hating"
- In contrast, hydrophilic – water "loving"

Amphiphilic Molecules in Water

- Nonpolar tails (hydrophobic portion) point away from water
- Polar heads (hydrophilic portion) are exposed to water
- Different amphipathic molecules aggregate in different ways based on the number of tails and size of the polar head group
- Micelles form when there is only 1 tail and take spherical form in aqueous solutions

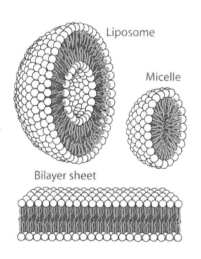

Chapter 5 – Chemical Kinetics

Introduction

- Reactions occur at different rates
    - Some are very quick, some are very slow, and many fall somewhere in between
- Knowing the rate of a reaction helps chemists plan out experiments and plan reactions accordingly
    - If you understand what contributes to rate, you can exert some control over a reaction
- Chemical equations (e.g., $Al_2O_3 \rightarrow Al + O_2$) don't tell you anything about how fast the reaction occurs
    - Some reactions occur in a series of smaller steps that lead to the final product

Reaction Rates

Generally, rates are defined as the change of something divided by change in time. This is true of reaction rates as well.

- Rate of a reaction can be written with respect to any compound in that reaction
    - But, there can only be one numerical value for a rate of reaction
- If you plot average rate data as a function of time, you will see that the reaction rate constantly changes (consider the graph below)

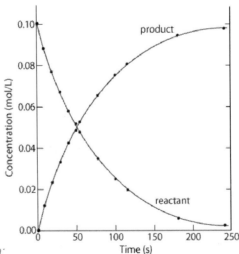

Rate Data for a Reaction:

- As you might notice, rate depends on the concentration of the reactants

- Since the rate of a reaction is effected by the concentration of the reactants we can write mathematical relationships linking the concentration of reactants with how fast the reaction occurs (i.e. we can write rate laws)
- General Reaction Rates
    - Consider the general chemical equation: aA + bB → cC + dD
    - Rate for this reaction would be defined as:
        - $$Rate = -\frac{1}{a}\left(\frac{\Delta[A]}{\Delta t}\right) = -\frac{1}{b}\left(\frac{\Delta[B]}{\Delta t}\right) = \frac{1}{c}\left(\frac{\Delta[C]}{\Delta t}\right) = -\frac{1}{d}\left(\frac{\Delta[D]}{\Delta t}\right)$$
- A simple rate law example:
    - Consider the decomposition reaction where: A → products
    - If the reverse reaction is negligible, then the rate law is: Rate = k[A]^n
        - k is called the **rate constant**
        - n is called the **reaction order**

Reaction Orders

- Reaction order (denoted as "n") determines how the rate depends on the concentration of the reactant
    - n = 0, zero order, rate is independent of [A]
    - n = 1, first order, rate is directly proportional to [A]
    - n = 2, second order, rate is proportional to the square of [A]
- Each order results in a different type of curve when graphed

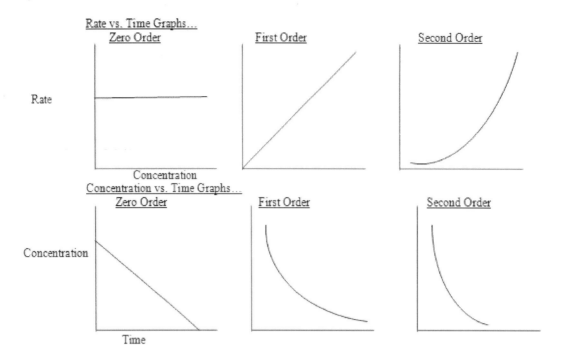

- ***Important*:** You can only determine reaction orders through an experiment
    - Reaction orders are **not** related to the stoichiometry of a reaction

Steps for Finding Rate Law

- Pick two solutions where one reactant stays same, but another changes
- Write the rate law for both using as much information as you have
- Form a ratio from the two and solve for an order
- Repeat the 3 steps above for another pair of solutions
- Use any reaction to get the value of k

An Example Problem

- Imagine we are considering the general reaction: A + B → products
- And that we determined the following information from an experiment:

Exp.	Initial A (mmol/L)	Initial B (mmol/L)	Init. Rate of Formation of products (mM min^{-1})
1	4.0	6.0	1.60
2	2.0	6.0	0.80
3	4.0	3.0	0.40

- Look at experiments 1 and 2
    - From experiment 2 to 1, we see that the concentration of A doubles (while B is held constant) and the rate also doubles
    - Doubling of the rate with a doubling of the concentration shows that the reaction is first order with respect to A
- Next look at experiments 1 and 3
    - Concentration of B is halved (while A is held constant)
    - When B is halved, the overall rate drops by a factor of 4 (which is the square of 2)
    - This shows the reaction is second order with respect to B
- The rate law would be written as: rate = k [A] [B]^2
- You can use any reaction to get the value of k (I will use experiment 1)
    - Rate = k [A] [B]^2
    - 1.60 mM min^{-1} = k (4.0 mM) (6.0 mM)^2
    - k = 0.011 mM^{-2} min^{-2}

Integrated Rate Law and Half Life Formulas

Order	Integrated Rate Law	½ Life
Zero	$[A]_t = kt + [A]_o$	$\dfrac{[A]_o}{2k} = t_{1/2}$
First	$\ln([A]_t) = -kt + \ln([A]_o)$	$t_{1/2} = \dfrac{0.693}{k}$
Second	$\dfrac{1}{[A]_t} = kt + \dfrac{1}{[A]_o}$	$\dfrac{1}{k[A]_o} = t_{1/2}$

- Note that the integrated rate law equations are in the form y = mx + b
    - y = mx + b is the formula for a straight line
    - So, the plot of ln[A] vs. time for the reactions will yield a straight line
- For 2nd order, half-life depends on initial concentration
    - As concentration decreases, the half-life increases
- Half-life for zero order reactions depends on concentration as well

- However, notice that the half-life doesn't depend on reactant concentration for the 1st order reactions
    - Half-life for a 1st order reaction is constant

Temperature and Rate

- Generally, rates of reaction are sensitive to temperature
    - Rate = $k[A]^n$, so where do we factor in temperature?
        - It is reflected in the constant k
        - Generally, increasing temperature increases k

Chemistry of Catalysis

Catalysts do not change the direction of a chemical reaction and they have no effect on equilibrium!

- Function by lowering the activation energy, which speeds up the reaction
- As the reaction progresses; reactants become products
    - Depicted as a reaction coordinate diagram

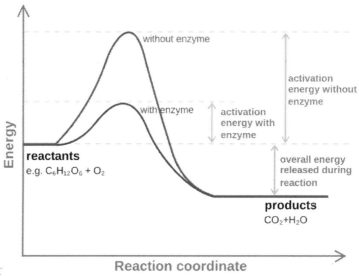

Reaction Coordinate Diagram:

    - Progress of the reaction is indicated on the x-axis
    - Free energy (G) is indicated on the y-axis
- Reactants pass through the transition state (‡) and become products
    - Enzymes increase the reaction rate by binding tightly to the transition state and stabilizing it

Spontaneous vs. Non-spontaneous Reactions

- Spontaneous if ΔG_{rxn} negative

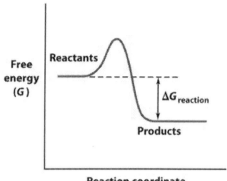

General Spontaneous Reaction Diagram:

- Non-spontaneous if ΔG_{rxn} positive

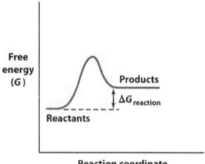

General Non-spontaneous Reaction Diagram:

- Enzymes (catalysts) lower the energy barrier
    - Make it easier to reach the transition state

Chapter 6 – Chemical Equilibrium

What is Equilibrium?

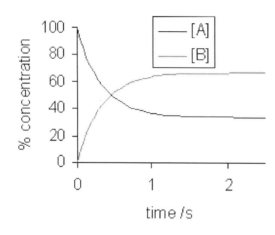

- As a system is approaching equilibrium, both the forward and reverse reactions are occurring at different rates

- Chemical equilibrium is established when a reaction and its reverse reaction occur at the same rate

- Once equilibrium is established, the amount of both the reactant and product remains constant

- In a system at equilibrium, both the forward and reverse reactions are running simultaneously so we write the chemical equation for the reaction with a double arrow

    - Example: $H_2 \rightleftharpoons 2H$

Equilibrium Constant

- Consider the reaction: $N_2O_4(g) \rightleftharpoons 2NO_2(g)$

    - Rate law for the forward reaction would be: Rate = k_f [N_2O_4]

    - Rate law for the reverse reaction would be: Rate = k_r [NO_2]^2

    - At equilibrium the two rates would be the same so we can rearrange the equations to get:

    $$K_{eq} = \frac{k_f}{k_r} = \frac{[NO_2]^2}{[N_2O_4]}$$

    - K_{eq} (aka K_c) is the equilibrium constant

- In general, the reaction: $aA + bB \rightleftharpoons cC + dD$

  o Results in the equilibrium expression: $K_c = \dfrac{[C]^c [D]^d}{[A]^a [B]^b}$

- Equilibrium can be reached from either the forward or reverse direction
    - o K_c, the final ratio of $[NO_2]^2$ to $[N_2O_4]$, will reach a constant no matter what the initial concentrations of NO_2 and N_2O_4 are (as long as time is held constant between them)
    - o Also note that the equilibrium constant of a reaction in the reverse reaction is the reciprocal of the equilibrium constant of the forward reaction

- If K >> 1, the reaction is said to be *product-favored*
    - o Products predominate at equilibrium

- If K << 1, the reaction is said to be *reactant-favored*
    - o Reactants predominate at equilibrium

- If a reaction consists of many individual steps, you can add the equilibrium constants for the individual steps to determine the equilibrium constant for the entire reaction

- Pressure is proportional to concentration for gases, because of this, the equilibrium expression can also be written in terms of partial pressures (instead of concentration)

  o $K_p = \dfrac{(P_C)^c (P_D)^d}{(P_A)^a (P_B)^b}$

    - o K_p and K_c are related to one another by the equation: $K_p = K_c (RT)^{\Delta n}$
        - ▪ Δn = (moles of gaseous product) – (moles of gaseous reactant)

Homogeneous and Heterogeneous Equilibrium

A homogeneous equilibrium is an equilibrium where all reagents and products are found in the same phase (solid, liquid, or gas). A heterogeneous equilibrium is an equilibrium where they are in different phases.

- Concentrations of liquids and solids can be obtained by the following:

  $\dfrac{density}{molar\ mass} = \dfrac{g/L}{g/mol} = \dfrac{mol}{L}$

- Concentration of solids and liquids are not used to form an equilibrium expression
  - Consider the reaction: $PbCl_{2(s)} \rightleftharpoons Pb^{2+}_{(aq)} + 2\,Cl^-_{(aq)}$
  - The equilibrium constant for the reaction would be:
    - $K_c = [Pb^{2+}][Cl^-]^2$

Reaction Quotient (Q)

- To calculate Q, you have to substitute the initial concentrations of reactants and products into the equilibrium expression
- Q gives the same ratio as the equilibrium expression but for a system that is **not** at equilibrium
  - If Q = K, the system is at equilibrium
  - If Q > K, there is more product, and the equilibrium shifts to the reactants
  - If Q < K, there are more reactants, and the equilibrium shifts to the products
  - The shifting of equilibrium is Le Châtelier's Principle
    - Essentially the principle states that equilibrium position shifts to counteract the effect of a disturbance (change in temperature, concentration, etc.)

Chapter 7 – Acid Base Equilibrium

Definitions and Conventions

Acid-base reactions are a type of chemical process typified by the exchange of one or more hydrogen ions (i.e. exchange/transfer of a proton).

- Arrhenius definition
    - Acid – substance that produces H^+ ions in aqueous solution
        - We now know that H^+ reacts immediately with a water molecule to produce a hydronium ion (H_3O^+)
    - Base – substance that produces OH^- ions in aqueous solution
- Bronsted-Lowry definition
    - Acid – proton donor
    - Base – proton acceptor
    - Bronsted-Lowry definition does not require water as a reactant
- Conjugate acids and bases
    - Conjugate base – species that is formed when an acid donates a proton to a base
    - Conjugate acid – species that is formed when a base accepts a proton from an acid
    - Conjugate acid-base pair – pair of molecules or ions that can be interconverted through the transfer of a proton

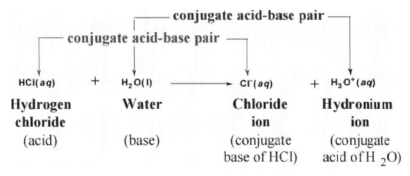

- Curved arrows are used to show the flow of electrons in an acid-base reaction

$$CH_3-C(=O)-O-H + :NH_3 \rightleftharpoons CH_3-C(=O)-O:^- + H-NH_3^+$$

Acetic acid (proton donor) + Ammonia (proton acceptor) ⇌ Acetate ion + Ammonium ion

- Neutralization is the reaction of an H⁺ (H_3O^+) ion from the acid and the OH^- ion from the base to form water, H_2O
    - Neutralization reaction is exothermic and releases approximately 56 kJ per mole of acid and base
- Determining acidic, basic, and neutral from concentration of H_3O^+ and OH^-
    - Neutral: $[H_3O^+] = [OH^-]$
    - Acidic: $[H_3O^+] > [OH^-]$
    - Basic: $[H_3O^+] < [OH^-]$

Strengths of Acids and Bases

- Strength of an acid is expressed by an equilibrium constant
    - Equilibrium expression for the dissociation of an uncharged acid (HA)

$$HA + H_2O \rightleftharpoons A^- + H_3O^+$$

$$K_{eq} = \frac{[H_3O^+][A^-]}{[HA][H_2O]}$$

    - K_a, the acid dissociation constant, is given by:

$$K_a = K_{eq}[H_2O] = \frac{[H_3O^+][A^-]}{[HA]}$$

        - This is because the concentration of water is high, and does not significantly change during the reaction, so its value is absorbed into the constant
        - The stronger the acid, the larger the K_a, and the more it will dissociate in solution

- Strong acids completely dissociate into ions in water
    - Strong acids are HI, HBr, $HClO_4$, HCl, $HClO_3$, H_2SO_4, and HNO_3
        - Their conjugate bases are weak

- Weak acids only partially dissociate into ions in water
- Polyprotic acids are acids that are capable of losing more than a single proton per molecule during an acid-base reaction
    - Phosphoric acid is a weak acid that normally only loses one proton but it will lose all three when reacted with a strong base at high temperatures
    - If the difference between the K_a for the first dissociation and subsequent K_a values is 10^3 or more, the pH generally depends **only** on the first dissociation
- In any acid-base reaction, the equilibrium favors the reaction that moves the proton to the stronger base
- The more polar the H-X bond and/or the weaker the H-X bond, the more acidic the compound
- Strong base – a base that is present almost entirely as ions (one of the ions is OH$^-$)
    - Strong bases are NaOH, KOH, LiOH, RbOH, CsOH, Ca(OH)$_2$, Ba(OH)$_2$, and Sr(OH)$_2$
- Weak base – a base that only partially ionizes in water
    - The general weak base reaction is written as: $\ddot{B} + H_2O \rightleftharpoons HB^+ + OH^-$
    - The equilibrium constant expression for this reaction is:
    $$K_c = K_b = \frac{[HB^+][OH^-]}{[B]}$$
        - K_b is called the base-dissociation constant
- K_a and K_b can be related to one another using the following formula:
    - $K_a \times K_b = K_w$
    - K_w is the ionization constant for water at 25 °C
        - K for water is: $K = \frac{[H^+][OH^-]}{[H_2O]}$ or $K_w = [H^+][OH^-] = 1.0 \times 10^{-14}$

Finding Concentration of Species in Solution from K_a

- Given: 0.10 M HNO_2 (nitrous acid), $K_a = 4.5 \times 10^{-4}$

- Set up a table to help you keep track of what is happening during the reaction

	HNO_2 ⇌	H^+ +	NO_2^-
Initial Concentration	0.10 M	0.00 M	0.00 M
Change in Concentration	-x	+x	+x
Equilibrium Amount	0.10 M - x	x	x

- Some of the reactant will become product, that is why the change in concentration is negative x

- We are forming some amount of product in this reaction so the change in concentration is positive x

- We now need to solve for x, first set up the K_a equation

    o $K_a = \dfrac{[H^+][NO_2^-]}{[HNO_2]} = \dfrac{[x][x]}{[0.10-x]}$

    o $4.5 \times 10^{-4} = \dfrac{x^2}{0.10-x}$

    o The simple approach:

    - It is accepted that as long as X < 5% of [HA], where [HA] = concentration of the acid, you can assume x is negligible and that (0.10 - x) = 0.10, making the K_a equation:

        - $4.5 \times 10^{-4} = \dfrac{x^2}{0.10}$

        - $4.5 \times 10^{-5} = x^2$

        - $6.7 \times 10^{-3} = x$

    o The exact approach (quadratic formula):

    - However, if you can't assume x is going to be smaller than 5% you have to set up the exact formula

    - The quadratic equation is: $ax^2 + bx + c = 0$

    - $4.5 \times 10^{-4} = \dfrac{x^2}{0.10-x}$ → $x^2 + 4.5 \times 10^{-4}x - 4.5 \times 10^{-5} = 0$

        - a = 1, b = 4.5×10^{-4}, and c = -4.5×10^{-5}

- The quadratic formula: $x = \frac{-b \pm \sqrt{b^2 - 4ac}}{2a}$
    - To solve, substitute in the values for a, b, and c
    - $x = \frac{-4.5 \times 10^{-4} \pm \sqrt{(4.5 \times 10^{-4})^2 - 4(1)(-4.5 \times 10^{-5})}}{2(1)}$
    - $x = 6.5 \times 10^{-3}$
        - Always use only the positive root, the negative root does not make sense in the context of these sort of problems
  - If you compare the answer from both approaches (6.7×10^{-3} vs. 6.5×10^{-3}) you can see that the answers are pretty much the same
    - However, remember that you can only use the simple approach if X< 5% of [HA]
- Since we found x, we only need to substitute the value into the "Equilibrium Amount" section of the table we set up earlier to find the concentration of species in solution
  - [HNO$_2$] = (0.10 M – x) = (0.10 M – 6.5 x 10^{-3}) = 0.0935 M
  - [H$^+$] = x = 6.5 x10^{-3} M
  - [NO$_2^-$] = x = 6.5 x10^{-3} M

pH and pOH

- pH is defined as the negative, base-10 logarithm of the hydronium ion concentration
  - pH = -log[H$_3$O$^+$] → [H$_3$O$^+$] = 10^{-pH}
- In pure water:
  - K_w = [H$_3$O$^+$] [OH$^-$] = 1.0 x 10^{-14}
  - Because in pure water [H$_3$O$^+$] = [OH$^-$];
    - [H$_3$O$^+$] = (1.0 × 10^{-14})$^{1/2}$ = 1.0 x 10^{-7}
    - pH = -log[1.0 x 10^{-7}] = 7.00
    - 7.00 is considered neutral pH

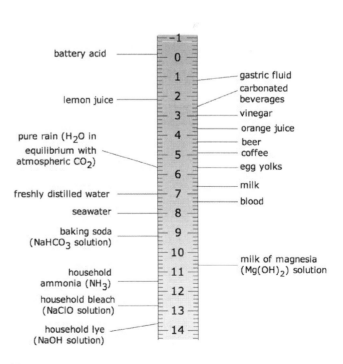

- An acid has a higher [H_3O^+] than pure water
    - pH is < 7
- A base has a lower [H_3O^+] than pure water
    - pH is > 7
- p in pH is a clue to take the negative log of the quantity, this is true for pOH and pK_w:
    - pOH = -log[OH^-] → [OH^-] = 10^{-pOH}
    - pK_w = - log(K_w) → K_w = 10^{-pK_w}

pK_a and Trends

- K_a = 10^{-pka}
- pK_a = -log(K_a)
    - Lower the pK_a, the stronger the acid
        - Higher the pK_a, the weaker the acid
    - Lower the pK_a, the weaker the conjugate base
        - Higher the pK_a, the stronger the conjugate base
- Equilibrium favors the side of the weakest acid and weakest base
    - Equilibrium favors the side with the highest pK_a
        - Thus, pK_a can be used to predict in which direction equilibrium lies

Percent Ionization Formulas

- $Percent\ ionization = \dfrac{amount\ ionized}{total\ in\ solution} \times 100\%$
- $Percent\ ionization = \dfrac{[A^-]}{[HA]+[A^-]} \times 100\%$

Reactions of Anions and Cations with Water

- Anions are bases
    - They can react with water in a hydrolysis reaction to form OH^- and the conjugate acid:
    $X^-(aq) + H_2O(l) \rightleftharpoons HX(aq) + OH^-(aq)$
- Cations with acidic protons (like NH_4^+) lower the pH of a solution because they release H^+ ions in solution

- Most metal cations that are hydrated in solution also lower the pH of the solution
    - Act by associating with H_2O and making it release H^+
- Attraction between nonbonding electrons on oxygen and the metal causes a shift of the electron density in water
    - This makes the O-H bond more polar and the water more acidic
    - Greater charge and smaller size make a cation more acidic

Effects of Cations and Anions

- An anion that is the conjugate base of a strong acid will not affect the pH
- An anion that is the conjugate base of a weak acid will increase the pH
- A cation that is the conjugate acid of a weak base will decrease the pH
- Cations of a strong Arrhenius base will not affect the pH
- Other metal ions will cause a decrease in pH
- When a solution contains both the conjugate base of a weak acid and the conjugate acid of a weak base, the effect on pH depends on the K_a and K_b values

CHAPTER 8 – SOLUBILITY EQUILIBRIUM

What is Solubility Equilibrium?

- If an "insoluble" or slightly soluble material is placed in water an equilibrium forms between the undissolved solids and ionic species in solutions
    - Solids continue to dissolve, while ion-pairs continue to form solids
    - The rate of dissolution is equal to the rate of precipitation
- Consider the reaction: $AgCl(s) \rightleftharpoons Ag^+(aq) + Cl^-(aq)$
    - $K = \dfrac{[Ag^+][Cl^-]}{[AgCl]}$
    - However, remember that since AgCl is a pure solid it isn't considered in K and thus the equation can be rewritten as: $K_{sp} = [Ag^+][Cl^-]$

Solubility Product (K_{sp})

- General expression: $M_mX_n(s) \rightleftharpoons mM^{n+}(aq) + nX^{m-}(aq)$
    - Solubility product for the general expression: $K_{sp} = [M^{n+}]^m[X^{m-}]^n$
- Example of how to find solubility (s) from K_{sp}:
    - $AgCl(s) \rightleftharpoons Ag^+(aq) + Cl^-(aq)$
        - $K_{sp} = [Ag^+][Cl^-] = 1.6 \times 10^{-10}$
    - If s is the solubility of AgCl, then:
        - $[Ag^+] = s$ and $[Cl^-] = s$
        - $K_{sp} = (s)(s) = s^2 = 1.6 \times 10^{-10}$
        - $s = 1.3 \times 10^{-5}$ mol/L
- Another example:
    - $Ag_3PO_4(s) \rightleftharpoons 3Ag^+(aq) + PO_4^{3-}(aq)$
        - $K_{sp} = [Ag^+]^3[PO_4^{3-}] = 1.8 \times 10^{-18}$
    - If the solubility of Ag_3PO_4 is s mol/L, then:
        - $K_{sp} = (3s)^3(s) = 27s^4 = 1.8 \times 10^{-18}$
        - $s = 1.6 \times 10^{-5}$ mol/L

Factors Affecting Solubility

- Temperature
    - Generally, solubility increases with temperature
- Common ion effect
    - Common ions reduce solubility
    - Consider the following solubility equilibrium:
        - $AgCl(s) \rightleftharpoons Ag^+(aq) + Cl^-(aq)$; $K_{sp} = 1.6 \times 10^{-10}$
        - The solubility of AgCl is 1.3×10^{-5} mol/L at 25 °C.
        - If NaCl is added, equilibrium shifts left due to increase in [Cl^-] and some AgCl will precipitate out
        - For example, if [Cl^-] = 1.0×10^{-2} M,
            - Solubility of AgCl = $(1.6 \times 10^{-10})/(1.0 \times 10^{-2})$ = 1.6×10^{-8} mol/L
- pH of solution
    - pH affects the solubility of ionic compounds in which the anions are conjugate bases of weak acids
    - Consider the following equilibrium:
        - $Ag_3PO_4(s) \rightleftharpoons 3Ag^+(aq) + PO_4^{3-}(aq)$;
        - If HNO_3 is added, the following reaction occurs:
            - $H_3O^+(aq) + PO_4^{3-}(aq) \rightleftharpoons HPO_4^{2-}(aq) + H_2O$
        - This reaction reduces PO_4^{3-} in solution, causing more solid Ag_3PO_4 to dissolve
- Formation of complex ion
    - Formation of complex ion increases solubility
    - Many transition metals ions have strong affinity for ligands to form complex ions
        - Ligands are molecules such as: H_2O, NH_3 and CO, or anions, such as F^-, CN^- and $S_2O_3^{2-}$
    - Complex ions are soluble
        - So the formation of complex ions increases solubility of slightly soluble ionic compounds

Predicting Formation of Precipitate

- $Q_{sp} = K_{sp}$
    - Saturated solution, but no precipitate
- $Q_{sp} > K_{sp}$
    - Saturated solution, with precipitate
- $Q_{sp} < K_{sp}$
    - Unsaturated solution
- Q_{sp} is ion product expressed in the same way as K_{sp} for a particular system
- Example Question: 20.0 mL of 0.025 M Pb(NO$_3$)$_2$ is added to 30.0 mL of 0.10 M NaCl. Predict if precipitate of PbCl$_2$ will form. Given: K_{sp} for PbCl$_2$ = 1.6 x 10^{-5}
    - [Pb^{2+}] = (20.0 mL x 0.025 M)/(50.0 mL) = 0.010 M
    - [Cl$^-$] = (30.0 mL x 0.10 M)/(50.0 mL) = 0.060 M
    - Q_{sp} = [Pb^{2+}][Cl$^-$]^2 = (0.010 M)(0.060 M)^2 = 3.6 x 10-5
    - $Q_{sp} > K_{sp}$, so PbCl$_2$ will precipitate

Chapter 9 – Electrochemistry

What is Electrochemistry?

Electrochemistry is a branch of chemistry concerned with the study of the relationship between electron flow and redox reactions.

- Oxidation-reduction reactions (aka redox reactions) – reactions that involve a partial or complete transfer of electrons from one reactant to another
    - Oxidation = loss of electrons
    - Reduction = gain of electrons
        - Trick for remembering which is which - OIL RIG
        - **OIL** - **O**xidation **I**s **L**osing electrons
        - **RIG** - **R**eduction **I**s **G**aining electrons
    - Oxidation and reduction always occur simultaneously

Oxidation Number = valence e^- – (unbonded e^- + bonding e^-)

- Oxidizing agent – species that oxidizes another species
    - It is itself reduced – gains electrons
- Reducing agent – species that reduces another species
    - It is itself oxidized – loses electrons

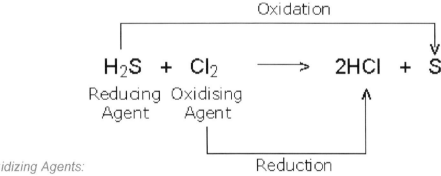

Reducing and Oxidizing Agents:

- Oxidation number (aka oxidation state) – actual charge an atom in a molecule would have if all the electrons it was sharing were transferred completely, not shared

Formal Charge vs. Oxidation Number

- Formal charges are used to examine resonance hybrid structures
    - Oxidation numbers are used to monitor redox reactions

- Formal Charge
    - Bonding electrons are assigned equally to the atoms
        - Each atom has half the electrons making up the bond
    - Formal Charge = valence e⁻ − (unbonded e⁻ + ½ bonding e⁻)
- Oxidation Number
    - Bonding electrons are transferred completely to the more electronegative atom

Half-Reactions

- Redox reactions can be written in terms of two half-reactions
    - One involves the loss of electrons (oxidation)
    - The other involves the gain of electrons (reduction)
    - Example: $Fe^{2+} + Ce^{4+} \rightarrow Fe^{3+} + Ce^{3+}$

$$Fe^{2+} \rightarrow Fe^{3+} + e^- \quad \text{(oxidation half-reaction)}$$
$$\underline{Ce^{4+} + e^- \rightarrow Ce^{3+} \quad \text{(reduction half-reaction)}}$$
$$Fe^{2+} + Ce^{4+} \rightarrow Fe^{3+} + Ce^{3+}$$

- A balanced redox equation has to have charge balance
    - Number of electrons lost in the oxidation half-reaction must be equal to the number of electrons gained in the reduction half-reaction

Rules for Balancing Redox Reactions

Balancing Redox Equations with Ion-Electron Method or Half-Reaction Method

- Write two half-reactions and balance both for:
    - The number of the key atom (i.e. the atom changing oxidation numbers)
    - Change in oxidation number with electrons
- Add half-reactions so electrons cancel
- Balance charge with OH⁻ (if the reaction is occurring in a base) or H⁺ (if the reaction is occurring in an acid)
- Balance O atoms with H_2O
- Check that there is no net change in charge or number of atoms

Oxidation Number Method

- Determine oxidation number of atoms to see which ones are changing

- Put in coefficients so that no net change in oxidation number occurs
- Balance the remaining atoms that are not involved in change of oxidation number
- Example: Consider the reaction: $HNO_3 + H_2S \rightarrow NO + S + H_2O$
    - Oxidation numbers: N = 5, S = -2 → N = 2, S = 0
        - N goes from 5 → 2
            - Δ = −3 reduction
        - S goes from -2 → 0
            - Δ = +2 oxidation
    - Multiply N by 2 and S by 3
        - $2\ HNO_3 + 3\ H_2S \rightarrow 2\ NO + 3\ S + H_2O$
    - Balance O in H_2O
        - 6 (Ox) → 2(Ox) + 4 H_2O
    - Write the final reaction and make sure it is balanced (same number of atoms on left and right side)
        - $2\ HNO_3 + 3\ H_2S \rightarrow 2\ NO + 3\ S + 4\ H_2O$

Voltaic (Galvanic) Cells

Voltaic cells are electrochemical cells in which a product-favored (spontaneous) redox reaction generates an electric current. The reaction produces an electron flow through an outside conductor (wire). Requirements for voltaic cells:

- Anode - an electrode (i.e. conductor such as metal strip or graphite) where oxidation occurs
- Cathode - an electrode where reduction occurs
- Salt bridge - tube of an electrolyte (sometimes in a gel) that is connected to the two half-cells of a voltaic cell
    - Salt bridge allows the flow of ions but prevents the mixing of the different solutions that would allow direct reaction of the cell reactants
    - Charge does not build up in half cells
        - Electrical neutrality must be maintained

Cell Diagrams

Cell diagrams are shorthand representation for an electrochemical cell.

- Anode is placed on the left side
- Cathode is placed on the right side
- Single vertical line represents a boundary between phases, such as between an electrode and a solution
- A double vertical line represents a salt bridge or porous barrier separating two half-cells

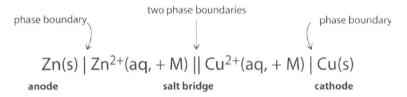

Electron Potential

- Electron flow in galvanic cell can do work/produce energy
- Electrical potential energy is measured in volts
  - 1 volt = (1 joule) / (1 coulomb)
    - Coulombs = amperes x seconds:
      - C = A x s   or   A = C / s

Standard Cell Voltages

- Cell voltages can be measured under standard conditions: 1 atm pressure, 25^0 C, and 1.0 M concentrations
  - Denoted as E^0_{cell}
- The standard cell potential is the sum of the standard potentials for the oxidative half-reaction and the reductive half-reaction
- If E^0_{cell} is positive, the net cell reaction is said to be product-favored (spontaneous)
- If E^0_{cell} is negative, the net cell reaction is said to be reactant-favored (nonspontaneous)

Standard Electrode Potentials

Standard electrode potentials are measured for half-reactions, relative to a standard hydrogen electrode potential (which has an assigned value of 0 volts).

- Each half reaction is written as a reduction
- Each half reaction could occur in either direction

- The more positive the standard electrode potential, the greater the tendency to undergo reduction
    - That means it is a good oxidizing agent
- The more negative the standard electrode potential, the greater the tendency to undergo oxidation
    - That means it is a good reducing agent
- If a half-reaction is written in the reverse direction, you must flip the sign of the corresponding standard electrode potential
- If a half-reaction is multiplied by a factor, the standard electrode potential is **not** multiplied by that factor

Cell Potential and Gibbs Free Energy

- Since a positive E^0_{cell} indicates a spontaneous reaction, you might imagine there is relationship between E^0_{cell} and free energy (ΔG^0)
- $\Delta G^0 = -nFE^0_{cell}$
    - n = # of moles of electrons transferred
    - F = Faraday constant = $9.65 \times 10^4 \frac{C}{mole \cdot e^-}$
    - **Important**: A positive E^0_{cell} would result in a negative ΔG^0, and a negative E^0_{cell} would result in a positive ΔG^0

Electrolytic Cells

Electrolytic cells consist of an electrolyte, its container, and two electrodes, in which the electrochemical reaction between the electrodes and the electrolyte produces an electric current.

- Properties of electrolytic cell
    - Requires energy (in the form of an electric current)
    - No physical separation is needed for the two electrode reactions
    - Usually no salt bridge is required
    - Conducting medium is molten salt or aqueous solution
    - For electrolytic redox reaction:
        - E^0_{cell} is negative
        - ΔG^0 is positive
        - K_c is small (<1)

Chapter 10 - Nuclear Chemistry

Radioactivity

Radioactivity is the emission of ionizing radiation or particles caused by the spontaneous disintegration of atomic nuclei.

- Types of radioactivity: alpha, beta, and gamma decay
    - Also positron emission
- Convention to be aware of:

Nuclear Equation

- Sum of the atomic numbers on both sides of the nuclear equation must be equal
- Sum of the mass numbers on both sides of a nuclear equation must be equal

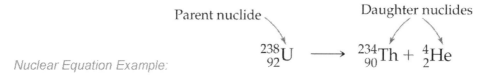

Nuclear Equation Example: $^{238}_{92}U \longrightarrow {}^{234}_{90}Th + {}^{4}_{2}He$

Alpha Decay

- Alpha particles are nuclear decay particles
    - An unstable nucleus emits a small piece of itself
- Alpha particles consist of two protons and two neutrons
- Alpha particle symbol: α
    - An α particle is a helium nucleus

Alpha Particle: $^{4}_{2}He \text{ or } {}^{4}_{2}\alpha$

- Alpha particles are ejected from the nucleus at a fairly low speed (approximately one-tenth the speed of light)
    - They are a minimal health risk to people unless ingested or inhaled

- Large mass nuclei tend to use alpha emission because it is a quick way for a large mass atom to lose a lot of nucleons (either a proton or neutron)

Alpha Decay Equation Example: $^{238}_{92}U \longrightarrow ^{234}_{90}Th + ^{4}_{2}He$

Beta Decay

- Beta radiation symbol: β or $^{0}_{-1}e$
- Beta emission is a nuclear decay process that ejects a high speed **electron** from an unstable nucleus
- Electron is formed within the nucleus by the breakdown of a neutron into a proton and electron
    - The electron is ejected from the system
    - The proton that was formed remains behind in the nucleus
        - Because of the addition of the proton, the atomic number of an element increases during beta emission
- Beta emission can be a significant health risk

Beta Decay Equation Example: $^{14}_{6}C \longrightarrow ^{14}_{7}N + ^{0}_{-1}\beta$

Gamma Decay

- Gamma radiation symbol: γ
- Gamma emission occurs primarily after the emission of a decay particle
- Gamma is a form of high energy electromagnetic radiation
    - It is a significant health risk
- After a particle is ejected from a nucleus the system may have some slight excess of energy, or exist in a meta-stable state
    - This slight excess of energy is released as gamma
- Gamma emission does not result in change of the isotope or the element
    - No mass and no charge change

Gamma Decay Equation Example: $^{125}_{53}I^{*} \longrightarrow ^{125}_{53}I + \gamma$

- The asterisk is used to represent that the element is in a high energy state

Positron Emission

- An unstable nucleus emits a positron
- A positron has the same mass as an electron but the charge is +1

Positron Emission Equation Example: $^{15}_{8}O \longrightarrow {}^{15}_{7}N + {}^{0}_{+1}\beta$

Half-life

Half-life is defined as the time for ½ of the parent nuclides to decay to daughter nuclides.

- All radioactive decay is first order
  - Rate = $-\frac{\Delta N}{\Delta t} = kN$
    - t – time
    - N - # of atoms
    - k = rate constant
  - $\ln(N_o/N) = k$
    - N_o - # of atoms at the starting time
- Half-life formula: $t_{\frac{1}{2}} = \frac{\ln(2)}{k} = \frac{0.693}{k}$
- Half-life is a constant

Carbon-14 Dating

- Carbon-14 dating can be used to date objects ranging from a few hundred years old to 50,000 years old
- Carbon-14 is produced in the atmosphere when neutrons from cosmic radiation react with nitrogen atoms
  - $^{14}_{7}N + {}^{1}_{0}n \rightarrow {}^{14}_{6}C + {}^{1}_{1}H$
- Living things take in carbon dioxide and have the same ^{14}C to ^{12}C ratio as the atmosphere
  - However, when a plant or animal dies, it stops taking in carbon as food or air
  - Radioactive decay of carbon starts to change the ratio of $^{14}C/^{12}C$
  - By measuring how much the ratio is lowered, we can determine how much time has passed since the plant or animal lived

- Half-life of cabon-14 is 5,720 years

Fission and Fusion

- In fission, a large mass nucleus is split into two or more smaller mass nuclei

$$^{235}_{92}U + ^{1}_{0}n \rightarrow ^{139}_{56}Ba + ^{94}_{36}Kr + 3\,^{1}_{0}n$$

Fission Equation Example:

- In fusion, small mass nuclei are combined to form a larger mass nucleus

$$^{2}_{1}H + ^{3}_{1}H \rightarrow ^{4}_{2}He + ^{1}_{0}n$$

Fusion Equation Example:

    o Fusion requires very high temperatures (in the millions of degrees) so that small nuclei can collide together at very high energies

CONCLUDING REMARKS

I hope this book has provided you tremendous value for your money and has helped you do better on your exams! If it has done both of these things, I have achieved my purpose in making this guide.

Furthermore, my goal is to create more books and guides that continue to deliver great value to readers like you for little monetary costs. Thank you again for purchasing this study guide and I wish you the best on your future endeavors!

- Dr. Holden Hemsworth

MORE BOOKS BY HOLDEN HEMSWORTH

DO YOU NEED HELP WITH OTHER CLASSES?

CHECK OUT OTHER BOOKS IN THE ACE! SERIES

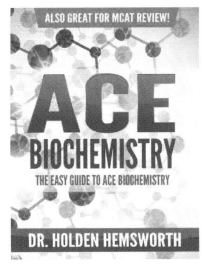

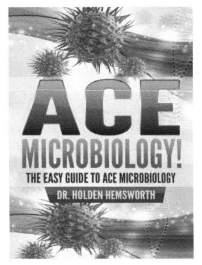

ALL BOOKS ARE LISTED ON MY AMAZON AUTHOR PAGE

MORE BOOKS COMING SOON!

Made in the USA
Lexington, KY
21 January 2019